Endangered Plants

Endangered Plants

D. M. Souza

Franklin Watts
A Division of Scholastic Inc.
New York • Toronto • London • Auckland • Sydney
Mexico City • New Delhi • Hong Kong
Danbury, Connecticut

Note to readers: Definitions for words in **bold** can be found in the Glossary at the back of this book.

Photographs © 2003: Corbis Images: 47 (David Butow), 32 (Jon Feingersh), 29 (Raymond Gehman), 15 top (Bob Rowan/Progressive Images), 23 (Richard Hamilton Smith); Dean Wm. Taylor: 16; National Geographic Image Collection/Bill Ballenberg: 15 bottom; Peter Arnold Inc.: 25 (Jeff & Alexa Henry), 26 (Alan Majchrowicz), 10 (Ray Pfortner), 30 (Julia Sims), 5 left, 13 (Still Pictures), 11 (Tom Vezo); Photo Researchers, NY: 39 top (Michael P. Gadomski), 42 (F. Gohier), cover (Jim W. Grace), 6 (Gilbert S. Grant), 51 (Jacques Jangoux), 12 (Maslowski Photo), 2 (Gary Meszaros), 18 (Jim Steinberg), 9 (Brenda Tharp); Smithsonian Institution, Washington, DC/Robert Hobdy: 34; Susan Cordell, Research Ecologist, USDA Forest Service/Institute of Pacific Islands Forestry: 35; Visuals Unlimited: 44 (D.Q. Cavagnaro), 22, 39 bottom (John D. Cunningham), 31 (Adam Jones), 48 (Link), 21 (NRCS), 41 (David Sieren), 20, 24 (John Sohlden), 49 (Inga Spence), 5 right, 36 (William J. Weber).

The photograph opposite the title page shows a prairie white fringed orchid.

Library of Congress Cataloging-in-Publication Data

Souza, D. M. (Dorothy M.)
    Endangered plants / D.M. Souza.
        p. cm. — (Watts library)
    Summary: Looks at various rare and endangered plants, discusses how they became endangered and efforts being made to protect them.
    Includes bibliographical references (p.   ).
    ISBN 0-531-12212-3 (lib. bdg.)   0-531-16248-6 (pbk.)
    1. Endangered plants—Juvenile literature. 2. Plant conservation—Juvenile literature. [1. Endangered plants. 2. Rare plants. 3. Plant conservation.] I. Title. II. Series.
QK86.A1 .S68 2003
581.68—dc21

2002015337

Contents

Chapter One
Why Care About Plants? 7

Chapter Two
Patches of Prairie 19

Chapter Three
Skeleton Forests 27

Chapter Four
Shrinking Wetlands 37

Chapter Five
Desert Under Attack 45

53 **Glossary**

56 **To Find Out More**

59 **A Note on Sources**

61 **Index**

Scientists believe that the Franklinia tree (blossom shown here) may date to the last ice age.

Why Care About Plants?

In 1765, John Bartram, a pioneer botanist, and his son, botanist and explorer William Bartram, found a 30-foot-tall (9-meter) tree with large fragrant white flowers growing along the banks of the Altamaha River in Georgia. They named the tree Franklinia (*Franklinia altamaha*) after Benjamin Franklin. A few years later, they returned to the site and collected seeds from the tree, but it was the last time they would see it growing in the wild. By 1776, the land where it had been growing had been cleared.

In 1987, a team of scientists collected plant samples from a forest on the island of Borneo. One sample taken from a small tree was later found to contain a valuable substance. When tested in the laboratory, it proved effective against the AIDS virus. While it did not cure the disease, it stopped the development of symptoms in HIV-positive patients. Anxious to obtain additional samples, the scientists returned to the forest but found that the lone tree had been cut for fuel or building materials. No other trees like it were on the island.

Each year, thousands of plant species disappear—some even before they can be discovered. Yet few of us are as concerned about their loss as we are about declining numbers of giant pandas, mountain gorillas, or Sumatran rhinos. If, however, we consider the role plants play in the lives of all creatures, perhaps we will better appreciate their importance and the necessity of saving them.

Powering the Planet

While making their own food by the process of **photosynthesis**, plants give off the oxygen that living organisms need. Their leaves release water vapor into the atmosphere, feeding cloud formations that affect weather. On a warm summer day, for example, a grove of trees may **emit** more than 8,000 gallons (30,320 liters) of water that cools the surrounding air.

Preserved

Franklinia was the first known native North American flowering plant to become **extinct** in the wild. Thanks to John and William Bartram and the seeds they saved, three cultivated Franklinias are growing outside the National Museum of Natural History in Washington, D.C.

Plants also nourish millions of creatures. Even meat eaters depend on plants because some of their prey are plant eaters. Ninety percent of the food we eat is supplied by only about 20 of the approximately 272,000 species of plants. Thousands of other edible plants, if cultivated, could help feed hungry people around the globe.

A small number of the known species of plants have been studied for their medicinal value. Some contain substances used in over-the-counter drugs. Others might hold chemicals

Vapor from the leaves of trees cools the surrounding air and feeds cloud formations.

Potent Plants

Novocain, the trademark name of the drug commonly used by dentists to relieve pain, originally came from coca leaves growing wild in the South American Andes Mountains. Taxol, used to treat ovarian cancer, comes from the bark of the Pacific yew tree (right), and the painkiller aspirin owes its origin to the willow tree.

that could be effective in treating various human illnesses and diseases.

Relatives of plants that died millions of years ago have become the coal and other fossil fuels that we depend on for energy. Many plants play a role in providing shelters and clothing and in the manufacture of various products. Milkweed, for example, was first used by Thomas Edison to produce natural rubber, and during World War II the fluffy "parachutes" attached to its seeds were turned into stuffing for life vests. The jojoba, a 10-foot-tall (3-m) shrub growing in

the deserts of the Southwest, yields oil similar to sperm whale oil. It is currently used to make soaps, shampoos, and creams. Yes, plants are an important part of our daily lives and that of the planet.

During World War II the fluffy parts of milkweed seeds like these were used to stuff life vests.

Linked Together

Each organism on Earth depends on a host of others for its survival. In forests, for example, the gray squirrel needs trees for shelter, food gathering, and nesting sites. The trees depend on insects and birds to pollinate their flowers and spread their

Many animals depend on trees for food, shelter, and nesting sites.

seeds. The pollinators need plants for food, shelter, and raising their young.

All of these living organisms also interact with nonliving parts of the environment such as water, soil, and climate. They form **ecosystems**, and millions of ecosystems make up a global network. Each species and each ecosystem has a role to play in the health of the entire network.

Unfortunately, these systems are vanishing at an alarming rate and with them the species they contain. In the United States alone, close to 5,000 native plants are under some degree of stress, and when one disappears it often sets off a chain reaction. The U.S. Fish and Wildlife Service recently noted that the loss of a single plant could easily affect up to thirty insects, plants, and higher animal species around it.

The Exterminators

At the beginning of the 20th century about 1.6 billion people inhabited Earth. Today that population has skyrocketed to more than 6 billion, causing our species to gradually crowd out all others. We are plowing fields of grasses to make room

for crops or pastures. We are cutting down forests and draining wetlands to build homes, highways, and shopping malls. Our activities are destroying ecosystems around the globe.

Extinctions are not new. Species have appeared and disappeared throughout Earth's history. Not only species, but also

The loss of forest around the globe affects millions of creatures.

Naming Species

In 2000, a group of scientists and businesspeople came together with the ambitious plan of inventorying all species on Earth within the next twenty-five years before those species disappear. As part of the ALL Species Foundation, scientists will use cutting-edge technology to locate undiscovered species in all areas of the globe, from coral reefs to rain forests. They hope to enter their findings in a database that everyone will be able to access.

larger groups of related organisms have vanished. Some past extinctions have been caused by dramatic changes in climate. Predators, insects, and diseases have brought about others. But most past extinctions were natural, and took thousands of years to complete. What is taking place today is unnatural and swift, and the cause is different. We are the exterminators.

Taking Action

During the past several decades, nations have taken steps to protect plants and animals in danger of disappearing. In 1973, the U.S. Congress passed the Endangered Species Act. Administered by the U.S. Fish and Wildlife Service and the National Marine Fisheries Service, the Endangered Species Program gathers information from scientists and conservationists to determine which organisms need help. If there is evidence that a species will become extinct if not given protection, it is classified as **endangered**. If it appears that without protection a species will be in trouble, it is listed as **threatened**, or likely to become endangered in the future.

Both foreign and native species are covered under the program. Those listed cannot be exported, imported, sold without permits, or removed from federal lands. Anyone found guilty of violations can be fined up to $100,000 and sentenced to a year in jail.

Individual states also have enacted laws to protect plants. Some states, such as Hawaii, California, and Florida, lead the nation in the number of endangered and threatened plants

It's important to make everyone aware of the habitats of endangered plants so that these plants can be protected.

Botanical gardens now harbor many threatened plants.

Comeback Kid

In 1848, Asa Gray, a Harvard botanist, classified a plant he found growing in the moist meadows of Shasta Valley in northern California. He named it the Shasta owl's clover (*Orthocarpus pachystachyus*). The plant was collected again in 1913, but after that it could not be located. It was presumed extinct until 1996, when a scientist found it, not in the moist meadows of the Shasta valley but on a higher, drier hillside. Because only eight plants were uncovered, the Shasta owl's clover remains on the endangered list.

found within their borders. Species in jeopardy are frequently stored or grown in botanical gardens to ensure their safety. Special attention is given to **endemics**, plants growing in only one specific area and found nowhere else. Nearly 90 percent of Hawaii's and 36 percent of California's native species are endemic.

Many conservation groups are also working to save plants. Occasionally, thanks to the efforts of these groups, species make a comeback and can be removed from the endangered or threatened lists. In a few cases, plants believed to be extinct have been rediscovered.

Although progress has been made, the outlook for many plants does not look promising. Each day, human activities continue to threaten more and more ecosystems. Unless we do something to protect them, the species they hold will disappear. Let's now take a closer look at a few ecosystems, highlight how they became threatened or endangered, and discover what some people are doing to save them and the plants they contain.

The central part of our nation was once a huge stretch of grasses and wildflowers.

Chapter Two

Patches of Prairie

When early settlers began their westward movement, the land between the eastern forests and the Rocky Mountains was an unending stretch of vegetation. Millions of acres of grasses and wildflowers known as **forbs** flourished and rippled in the winds like waves in a sea. Few trees grew except along the banks of rivers, and numerous birds, mammals, insects, and microorganisms populated the area. French traders and trappers

Big bluestem is a tall, sod-forming grass that grows from April to August.

who saw this vast expanse called it a **prairie**, a word meaning "meadow."

Unique Plants

Grasses occupy between 30 and 40 percent of the planet's land surface and are among the most successful plants on Earth. On the American prairie, the shortest varieties are in the dry western plains, the tallest in the wetter eastern region. Some tallgrasses, such as big bluestem, Indiangrass, and switchgrass, sometimes grow so tall that they can conceal a rider on horseback.

With their hollow stems, narrow leaves, and small, inconspicuous flowers, grasses grow from points beneath the soil, not from aboveground tips. This method of growth makes it easier for grasses to recover after being grazed, damaged by temperature extremes, or burned by fire. To survive droughts and drying winds, the plants plunge their

Other Names

Grasslands are found around the world. In Argentina they are known as pampas, in Africa, veldts, and in Asia, steppes.

roots deep into the soil in search of water. Some can reach as far as 11 feet (3.3m) below the surface.

Before the settlement of the prairie, grazing, fires, and fungi played important roles in the growth and spread of grasses and wildflowers. At one time, millions of bison, elk, deer, and bighorn sheep roamed the prairie, spreading nutrients through their urine and feces. Their hooves broke the soil and provided open spaces where new grass seeds and forbs could take root.

Fires were frequent events on the prairies, often started by lightning or by Native Americans trying to encircle bison during a hunt. These blazes burned dead plant debris and allowed sun and rain to **penetrate** to the soil. They released minerals that fed the next generation of grasses and increased

Largest Native Land Animals

More than 60 million bison were slaughtered in the early days of our nation. In 1990, fifty-one Native American tribes joined together to protect the remaining animals.

Prairie fires once provided nutrients that grasses and wildflowers needed.

The roots of many plants team up with fungi in a relationship called mycorrhiza, *which means "fungus root."*

the number of shoots and flowering stems. Fires also killed woody shrubs and trees that continually competed with the grasses.

Certain **fungi** that live in the soil still aid prairie plants, as they do many other species of plants. In a relationship known as **mycorrhiza**, the fungi penetrate the roots of the grasses and wildflowers, helping them take in nutrients and resist diseases. In return, the plants provide food for the fungi.

Changing Landscape

Settlers who arrived on the prairie found it a difficult place to carve out a life for their families. The lack of timber and firewood, intense summer heat, drying winds, and frequent fires were all challenging. With their simple tools and equipment, farmers struggled to cut through the thick roots of the vegetation and turn over the soil.

Not until 1837 and John Deere's invention of the steel-bladed plow was there a welcomed change. The plow allowed settlers to work the land and plant crops on a large scale. Prairie soil was suddenly discovered to be one of the most fertile in the world.

New settlers arrived, built more housing, and plowed new tracts of land. Railroad companies laid tracks and brought additional changes. Soon cities, towns, roads, and highways dotted the once vast prairie and greatly reduced the great herds of wild animals as well as the numbers of native peoples who depended on them.

Today, less than 1 percent of the tallgrass prairie remains, and most of it grows in graveyards and along roadsides. Hundreds of wildflowers are now listed as either threatened or endangered. Many small creatures that were a vital part of this ecosystem have also disappeared.

Where grasslands once grew, wheat, corn, and other grains now carpet the land.

Preserved

Today, while the prairie fringed orchid is extinct in Oklahoma and South Dakota, small populations are being preserved in some other midwestern and western states and in Manitoba, Canada.

Prairie Orchids

The white feathery blooms of the eastern and western prairie fringed orchids (*Platanthera leucophaea* and *P. praeclara*). once visible everywhere, are now found only in limited numbers. Some of these **perennials** grow as tall as 4 feet (1.2 m) with five thick, long leaves branching from a spindly stem. Mature orchids bloom for up to three weeks and may have as many as two dozen flowers on each plant. Pollinated by prairie hawkmoths, the flowers form seeds that in time produce a new generation of orchids.

Plowing destroyed many of these orchids. Applications of fertilizers and pesticides killed others or the fungi and pollinators they depended on. In 1989, the prairie fringed orchid was officially listed in both the United States and Canada as either threatened or endangered.

Reseeding Prairies

Concerned people, realizing the benefits that prairie plants once provided—erosion control, water purification, wildlife habitat—are attempting to restore them. They are exploring old cemeteries and searching for seeds of rare grasses and flowers. Some are planting these seeds on portions of their property. Native Americans are preserving grasslands on their reservations.

The Bureau of Land Management, U.S.D.A. Forest Service, and the National Park Service are working to reestablish grasslands and native wildlife on acreage they manage. Private organizations such as the Nature Conservancy, based in Virginia, have also purchased tracts of land where prairie grasses are being planted and preserved. As a result of these efforts, it is hoped that wildlife will one day return and future generations will be able to experience a little of the grandeur that was once the American prairie.

Human activities threaten ecosystems such as this one in Yellowstone National Park.

Ancient forests of the Pacific Northwest have been reduced by almost 90 percent.

Chapter Three

Skeleton Forests

We hear more about the destruction of tropical rain forests around the world than about the disappearance of native forests in the United States. Few people realize that at one time almost half the land of this nation was covered with trees, and that after the arrival of Europeans these ancient forests began disappearing. Colonists considered forestlands unproductive and cut trees to make room for crops. They used timber for building, furniture making, and heating, and what was not needed locally they shipped abroad.

Additional forests were leveled as people moved west in the 19th century. By the early part of the 20th century, only 13 percent of this country's original timberlands were still standing. Today less than 5 percent remain—mere skeletons of ancient forests.

Exotic insects are attacking stands of spruce-fir in southern Appalachia. Alien invaders are destroying native oaks in California and Oregon. Grazing livestock and the construction of dams have reduced cottonwoods, willows, and other trees once abundant along southwestern streams and rivers. Development, agriculture, and exotics have all but eliminated the dry forests of Hawaii.

Trees are essential for pure air and clean water supplies. They hold soils in place and prevent flooding and landslides. Forests provide habitat for a variety of other plants and animals. When they are destroyed, ecosystems suffer.

The Longleaf Pine

The longleaf pine (*Pinus palustris*) is unique in many ways. When young, it resembles a clump of grass rather than a tree. For the first few years of its life it does not grow much in height, but develops the strong **taproot** critical to its ability to survive fires. Fires raging through woodlands can level everything in their path. But for tens of thousands of years the fires kept longleaf pine forests healthy because the trees were able

Opposite:
Longleaf pine seedlings stay in the so-called "grass stage" for four to six years.

28

Endangered Bird

The red-cockaded woodpecker uses its sharp beak to hammer nesting holes in the trunks of **old-growth** longleaf pines. Their pecking causes sap to build up around the holes or run down the trunks and discourages prey, such as snakes, from reaching the nest. Decreasing old-growth forests have resulted in the birds' decline. Red-cockaded woodpeckers are now listed as endangered.

to withstand the fires that destroyed other plants competing for space.

Once the tree starts to shoot upward, it grows 4 to 6 feet (1.2 to 1.8 m) a year and may eventually reach 100 feet (30 m) in height. It has few side branches, a small open **crown**, and the longest needles of all southern pines. Numerous animals thrive wherever it grows.

Starting in May, the areas where the trees grow are hit by thunderstorms that peak in July and August. From time to time, lightning ignites the pine needles that have built up on the ground and sends a blaze raging across the land. Wiregrass (*Aristida beyrichiana*) and other surrounding plants only add fuel to the fire.

The thick bark of mature longleaf pines is fire resistant. Massive taproots protect the trees, while long needles direct fire away from tender buds. Although the aboveground parts of nearby perennials are consumed, their roots are not harmed and the plants soon resprout.

Colonists cut thousands of acres of longleaf pine forests and converted the land to agriculture. In time, farms, citrus orchards, strip mines, and residential and commercial developments dominated the land, and lightning fires were controlled to protect what had been built. As many as 90 million acres (36 million hectares) from Virginia to Texas that were once longleaf pine ecosystems have been reduced to less than 3 percent. More than thirty plant and animal species that were a part of this ecosystem are now either threatened or endangered.

A New Resource

For hundreds of years, fire kept longleaf pine forests open and parklike and allowed a variety of wildflowers and **legumes** to grow beneath the trees. When colonists arrived, they began using the trees. The tall, straight timbers were used to build homes, public buildings, and ships. Tar, a by-product, helped lubricate the axles of wagons and dress the wounds of livestock. Later, turpentine and tar were extracted and sold commercially.

These young students are helping to restore forests and ecosystems.

Getting Involved

Conservationists, private landowners, and forest managers are working together in an attempt to restore the longleaf pine forests. Thousands of seedlings have been planted in various states and their survival rate is being monitored. Many other groups are participating in efforts to revive the forests.

In 1999, almost one hundred seedlings were donated to students at an elementary school in King George County, Virginia. After months of careful attention, the seedlings, along with native Virginia yellow pitcher plants (*Sarracenia flava*), were planted along a highway right-of-way. At the end of one year, 61 percent of the plants had survived. The success of this project has inspired other students in other states to begin restoration programs of their own.

Another group, the Longleaf Alliance in Alabama, has developed programs that promote the values of the trees. Individuals, landowners, organizations, industries, and government agencies are becoming involved in reseeding projects. With the cooperation of everyone, perhaps the decline of the longleaf pine ecosystems can be reversed.

Dry Forests

The Hawaiian Islands are famous for their lush rain forests. Equally valuable are their dry forests that often receive less than 20 inches (51 centimeters) of rainfall each year. In these dry areas, trees such as the Wiliwili (*Erythrina sandwicensis*) and a variety of other plants, once thrived and blanketed the **leeward** mountain slopes of all the Hawaiian islands. While 40 percent of Hawaii's rain forests have now disappeared, 90 percent of its dry forests have become victims of development, agriculture, and invasion by exotics.

The destruction of the dry forests began with the first human arrivals. Early inhabitants cut trees and burned the forests to make way for crops. Europeans later arrived and planted highly flammable alien grasses that, when ignited by lava flows, killed the trees. Livestock and **feral** mammals consumed or trampled native vegetation. When lowlands were converted to sugarcane and pineapple plantations, the forests were further reduced.

Rarities

One of the rarest native trees endemic to the dry forests of Hawaii is the Hawaiian tree cotton (*Kokia drynarioides*). It is a member of the hibiscus family, and its bark was once used to

Useful Timber

The soft, light wood of the Wiliwili was once popular for building the outriggers of Hawaiian canoes, fishnet floats, and surfboards, and the red seeds of the tree were used to make leis.

The Hawaiian tree cotton with its colorful blooms is now increasingly rare in the wild.

make red dye for fishnets. As early as 1916, scientists noticed that the plant's habitat was in trouble, and they collected seeds and sent them to botanical gardens around the world. By 1984, *Kokia* was listed as an endangered species and currently fewer than ten specimens remain in the wild.

Another rare tree of the dry forests is Kauila (*Colubrina oppositifolia*). At one time, its hard wood was used for mallets, poles, fishing spears, and weapons. The wood was so strong that it could be substituted for metal.

Gradually, the tree was threatened by loss of habitat, foraging by goats and cattle, fires fueled by alien grasses, and attacks by insects, such as the black twig borer. In 1994 a group of citizens took action to have it protected. They argued that the preservation of Hawaii's dry forests was essential if these and other rare plants were to be saved. Kauila is now listed as endangered.

Islands of Hope

Since 1995, groups of scientists and local residents have been working together to restore dry forest ecosystems on the main island of Hawaii. On a 6-acre (2.4-ha) fenced tract of land, known as the Kaupulehu Forest Reserve, a kind of Noah's ark for plants has been created. The fence, in place for the past forty years, has kept out goats, cattle, and other animals that graze in the area. Workers have whacked weeds and uprooted alien fountain grass (*Pennisetum setaceum*) that once invaded the site, reducing the threat of fires. They have planted hundreds of seedlings of endangered dry forest plants and are closely watching them. If successful, the project could become a model for restoration of dry forests on other islands.

The Kaupulehu Forest Reserve has become a kind of Noah's ark for rare and endangered species.

Mangrove trees grow in wetlands.

Chapter Four

Shrinking Wetlands

Imagine an area more than four times the size of New York State. That is the size of the wetland acreage that once existed in the lower forty-eight states. About half of it has been filled, drained, or dredged for agriculture, logging, dam construction, oil drilling, and development.

Remnants of wetlands can be found in a variety of sizes and shapes. The vast Alaskan coastal marshes, the mangrove forests in Florida, numerous swamps, bogs, fens, and tiny potholes on the prairies—all are wetlands. Some are saturated with water year round; others hold water only during spring rains. Some hold freshwater, others salt water.

All of these wetlands are valuable for several reasons. They provide food and shelter for a variety of creatures. They clean and filter drinking water, control flooding, and offer recreational benefits. Wetlands also support species of plants and animals found nowhere else.

Bog Bodies

In several European bogs, the well-preserved remains of ancient humans have been uncovered along with artifacts that give clues to how the people lived.

The Story of Bogs

Thousands of years ago, retreating glaciers left deep cavities in poorly drained land. These depressions filled with rainwater and became the birthplace of bogs. Slowly, plants such as sphagnum moss and grasslike sedges grew around the edges of the wetlands and eventually spread over their surface. Floating mats of soft, spongy vegetation formed where other plants could take root.

When the bog plants died, they fell to the bottom and became **peat**. In some areas, the layer of peat accumulated until it was more than 23 feet (7 m) thick. Frequently, it hid the bodies of creatures that once roamed Earth.

Most bogs in the United States are in the northeast where the climate is cold and wet, but bogs known as **pocosins** can also be found in the southeast. To grow crops in these places, one of the first things colonists did was dig ditches to drain away the water. The manufacture of pipe from fired clay in the 19th century made their job easier. Earthenware pipes, or tiles,

Waterlogged habitats known as bogs are often home to carnivorous plants.

Peat, found in wetlands, is in demand as a soil conditioner.

were laid in ditches so that water could be diverted into creeks and streams.

In some areas, peat was collected and sold as fuel or as a soil conditioner. Once their habitats were changed, bog plants could no longer grow. In some cases, ecosystems that had taken thousands of years to form were destroyed in a matter of days or weeks.

The Meat Eaters

Sphagnum and peat cause bogs to be low in nutrients and high in acid. **Carnivorous** plants such as pitcher plants (*Sarracenia*), and sundews (*Drosera*) are a few of the unique plants able to grow in these places. They make up for the lack of nutrients in the soil by capturing insects and other small organisms. As many as thirteen different species of meat-eating plants can live in a single bog.

In some places, pitcher plants form dense floating mats. With leaves shaped like pitchers and nectar that oozes from special glands, they attract insects looking for a meal. Once an insect lands, it slips on the plant's downward-pointing hairs, falls into a pool of water at the base of the pitcher, and drowns. Special chemicals and microorganisms help the plant slowly digest the food and absorb its nutrients.

Sundews have delicate leaves that sometimes glisten with drops of nectar. When insects arrive for a sip of the sweet liquid, they cannot escape. No matter how hard they struggle, they remain glued to the plant while their body fluids are slowly sucked out. Afterward, their remains are released to blow away on the slightest breeze.

The destruction of bogs and other wetlands has reduced the numbers of these and other meat-eating plants. The green pitcher plant (*Sarracenia oreophila*) and the mountain sweet pitcher (*Sarracenia rubra jonesii*) have been placed on the federal list of endangered plants. Other carnivorous plants, such as the Venus flytrap (*Dionaea muscipula*), have become extremely rare in parts of North and South Carolina.

The sweet scent of a pitcher plant attracts insects that frequently end up becoming the plant's dinner.

Restoring Wetlands

Wetlands are no longer considered useless swamps. Through the efforts of the federal, state, and local governments and various conservation groups, wetlands are now seen as valuable natural resources that must be preserved. In certain areas, laws have been passed to prevent drainage of wetlands and to prohibit mining, development, or other activities that would affect these important ecosystems.

Some scientists believe that unless action is taken immediately, the Everglades will be destroyed by 2015.

The Nature Conservancy is working to restore or preserve many tracts of wetlands. Another group, Meadowview Biological Research Station, is restoring pitcher plant bogs in Maryland and Virginia. Scientists are searching for new bogs

and growing seeds of pitcher plants that they plan to reintroduce into the wild.

In King County in the state of Washington, a class of students recently undertook the restoration of a wetland near their school. Cans, bottles, and other debris littered the area. Purple loosestrife (*Lythrum salicaria*), an exotic plant, was crowding out natives. Students worked in groups to remove the litter and dig up the loosestrife. They planted more than one hundred natives, then presented wetland educational programs for other students and their parents. Not only did they help restore the wetland, but they also increased understanding of the role of wetland ecosystems.

One of the most extensive rescues of a wetland was begun in the 1990s. Hundreds of scientists from state and federal agencies set out to restore the Florida Everglades. Funded by Congress, this group is attempting to repair the damage done over the years to a unique ecosystem.

Lost Habitats

More than 97 percent of carnivorous plant habitats in wetlands of the southeast have been lost.

Almost a million people visit Death Valley each year.

Chapter Five

Desert Under Attack

An arid sweep of land stretching from southern California into parts of Nevada, Arizona, and Utah is known as the Mojave Desert. It is a place of low mountains, gentle slopes, sand dunes, volcanic cones, and an immense flatland called Death Valley. Extreme day and night temperatures and little rainfall limit the kinds of plant species that can grow in this place. Those that manage to survive show amazing adaptations.

You might think that the harshness of the desert would protect it from invaders. Unfortunately, over the years, the Mojave has evolved into rangeland for cattle and sheep, sites for mining explorations, a military training center, and a playground for off-road vehicles. These and other activities have taken their toll on the rare plants and animals in this fragile ecosystem.

Cacti and Company

The Mojave is home to thousands of plant species, about two hundred of which are endemic. Cacti grow along with yuccas, creosote bushes, and sagebrushes. Trees are few except for the spiky Joshua (*Yucca brevifolia*) that takes root at higher elevations.

In 1997, one delicate plant, the Lane Mountain milkvetch (*Astragalus jaegerianus*), was listed as endangered. This member of the pea family grows on rocky, coarse sands where it almost always winds itself around low desert shrubs. Only 840 plants were found, and the majority of them are now threatened by the proposed expansion of Fort Irwin. As a result, conservation groups have banded together to protect not only the Lane Mountain milkvetch, but also the threat-

Low Point

The floor of Death Valley is 282 feet (85 m) below sea level and is the lowest point in the Western Hemisphere. In 1933, President Herbert Hoover designated the valley a national monument.

ened desert tortoise, whose habitat is within the proposed expansion area.

Joshua trees are a common sight in the Mojave Desert, especially at higher elevations.

The Invaders

Millions of years ago, a sea covered the Mojave Desert. Later, the land uplifted and volcanic activity shaped and reshaped the land. Wind, water, and sand made it a place of ever-changing landscapes.

Native Americans were the first to settle in the Mojave. In the valley and surrounding mountains, they found everything they needed for survival: springs, fruit, seeds, plants. But

First opened by the Southern Pacific in 1876, the torturous route from the Mojave Desert to Bakersfield, California, passes through the Tehachapi Mountains.

adventurers, seeking a shortcut to the goldfields, stumbled onto the desert in 1849 and ended forever the natives' way of life.

Miners and prospectors searching for silver, copper, and other metals began cutting down trees. They built towns, turned cattle loose to graze on the plants, and dug canals to bring water from the Colorado River. Two railroads were built, cities sprang up, and roads and highways soon crisscrossed the desert.

During World War II, part of the Mojave became a training ground for troops, tanks, and other military equipment. Today, Fort Irwin occupies an area larger than the state of Rhode Island and the army is attempting to expand it even further. Desert plants are frequently leveled by war games, and the soil, delicately knit together in many places by **lichens**, mosses, and other flowerless plants, is constantly disturbed.

Also invading the desert with each new wave of arrivals have been non-native grasses. They have filled in the spaces between native shrubs and plants and created fire hazards. During the past decade, thousands of acres in the Mojave have burned, fueled by these grasses. After a fire, the invaders grow back thicker than before. Scientists are concerned that native species will be unable to recover.

Off-road and all-terrain vehicles (ATV) are also destructive invaders of the desert. Each year, when cooler temperatures arrive, the sand dunes of the Mojave turn into a playground for ATV riders. As many as 200,000 people descend on the

All-terrain vehicles churn up sand and destroy many fragile plant species.

desert, crushing plants and creatures as they race across the dunes at breakneck speeds. The area where they play has become the most popular off-road site in the country. These and other human activities are placing numerous desert species in peril.

Saving an Ecosystem

To protect a portion of this rare desert, Congress, in 1976, set aside 25 million acres (10 million ha) of land from the Mexican border north to Death Valley. The space is called the California Desert Conservation Area (CDCA), and the Bureau of Land Management (BLM) oversees it. For years, conservation groups have been dissatisfied with the way the Bureau has been doing its job. Livestock grazing, mining, road construction, and off-road vehicle activities have been allowed to continue in the conservation area in spite of the threat they pose to the ecosystem.

In March 2000, the Center for Biological Diversity, the Sierra Club, and other environmental groups filed a lawsuit against the Bureau on behalf of endangered desert species. Private citizens were urged to write letters to their senators, to officials in the Fish and Wildlife Service, and to the BLM in support of the lawsuit. In the end, the Center for Biological Diversity won and the changes were sweeping.

As part of the settlement, a sand and gravel mine was closed and future mining was prohibited on more than 3 million acres (1.2 million ha). Livestock grazing was reduced or prohibited in certain areas. Thousands of acres were closed to off-road vehicles, and the BLM was instructed to closely oversee and protect endangered plants and animals.

The task now facing the BLM is huge. Desert plants grow slowly, and those that have been harmed will take years to

It can take centuries to re-grow tropical forests cut down by humans or destroyed by fire.

recover. Some scientists estimate it may take as many as two centuries, but with the public's cooperation the unique plants and animals of the Mojave can be saved.

Other ecosystems around the globe are also threatened. Perhaps the greatest damage is being done in tropical areas where most species of plants are found. Scientists estimate that if we continue to destroy ecosystems as we have in the past, at least one-fifth of the plants (and animals) now in existence will be gone by 2030, and about one-half will vanish by the end of the century.

Unavoidably, some plants will disappear no matter what we do. This is only natural, but we must make certain that we are not responsible for the disappearances. If enough people become more aware of plants and their value, join organizations aimed at restoring natural areas, and take part in preserving ecosystems, many extinctions can be halted. Future generations will then be able to enjoy the wonders of the plant world and help uncover more of the secrets it holds.

Glossary

carnivorous—meat-eating

crown—the branches, leaves, and twigs of a tree

ecosystem—a community of organisms and the environment in which it lives

emit—to send out

endangered—threatened with extinction

endemic—an organism found in a particular place and nowhere else

exotic—not native, alien

extinct—no longer existing

feral—wild, untamed

forb—a plant that is not a grass

fungus—a living nonphotosynthetic organism that frequently helps plant roots take up nutrients in the soil

leeward—the side toward which the wind blows

legume—a plant such as peas, beans, and alfalfa that has nitrogen-fixing bacteria in its roots

lichen—a combination of a fungus and an alga or cyanobacteria, growing as one

mycorrhiza—a mutually beneficial relationship between a fungus and the roots of a plant where fungi receive food from the plant and help plant roots take up soil nutrients

old-growth—more than 100 years old

peat—dead and decayed plant material found in damp or marshy areas

penetrate—to enter

perennial—a plant that lives for more than one season

photosynthesis—the process by which leaves, using energy from the Sun, change water and carbon dioxide into sugars

pocosin—a bog found in the southeastern United States

prairie—a large tract of grassland

taproot—a deep central root that helps anchor some trees and plants in the ground

threatened—endangered

To Find Out More

Books

Burton, John. *Atlas of Endangered Species*. New York: Macmillan Publishing Co., 1991.

Halpern, Robert R. *Green Planet Rescue, Saving the Earth's Endangered Plants*. Danbury, CT: Franklin Watts, 1993.

Landau, Elaine, *Endangered Plants*. Danbury, CT: Franklin Watts, 1992.

Patent, Dorothy Hinshaw. *Biodiversity*. New York: Clarion Books, 1996.

Videos

Bogs, Swamps, and Marshes. Educational Images, Ltd., 1999.

Extinction. Schlessinger Media, 1993.

The Last Flower. Smithsonian World Series, 1985.

The Last Show on Earth. Bullfrog Films, 1993.

Plants and People: A Beneficial Relationship. Schlessinger Media, 2000.

The Shaman's Apprentice. Bullfrog Films, 2001.

Organizations and Online Sites

Center for Plant Conservation (CPC)
P.O. Box 299
St. Louis, MO 63166-0299
http://www.mobot.org/CPC
This organization is dedicated to preserving and restoring rare native plants in the United States.

Defenders of Wildlife
1101 14th St. NW #1400
Washington, DC 20005
http://www.defenders.org/index.html
Protecting native plants and animals in their natural habitats is the goal of this organization.

National Wildlife Federation
8925 Leesburg Pike
Vienna, VA 22184
http://www.nwf.org
This group is working to protect biodiversity and endangered species through educational programs.

The Nature Conservancy
4245 N. Fairfax Drive, Suite 100
Arlington, VA 22203-1606
http://nature.org
This nonprofit group works to preserve wild places, especially those containing rare or endangered species.

The World Conservation Union
Rue Mauverney 28
1196 Gland
Switzerland
http://www.iucn.org
This union of scientists around the globe assists countries in preserving biodiversity and using their natural resources wisely.

A Note on Sources

A number of plants are included on the threatened and endangered species lists. Before deciding which ones to include in this book, I read the scientific report "Endangered Ecosystems in the United States." The authors, Reed F. Noss, Edward LaRoe, and J. Michael Scott, emphasized the importance of protecting ecosystems if we wish to save disappearing species. This was the focus I decided to take.

Extinction: The Causes and Consequences of the Disappearance of Species by Paul Ehrlich, *Imperiled Planet, Restoring Our Endangered Ecosystems* by Edward Goldsmith, and *The Future of Life* by E. O. Wilson were a few of the books that gave me additional facts on the subject. A search through pamphlets, journals, and scientific papers at a nearby university library provided a wealth of details on specific areas and the threatened or endangered plants in them. Web sites of the

Environmental Protection Agency, the U.S. Fish & Wildlife Service, and the Nature Conservancy were also helpful. Finally, a visit to the Mojave Desert with its rare plants and animals capped my research.

Index

Numbers in *italics* indicate illustrations.

Algae, 8
ALL Species Foundation, 13
All-terrain vehicles (ATV),
    49–50

Bark, 10
Bartram, John, 7, 8
Bartram, William, 7, 8
Big bluestem, 20
Bison, 21
Blackburn's sphinx moth, 34
Bogs, 38, *39*, 40–41, 42
Botanical gardens, *15*, 16
Bureau of Land
    Management (BLM),
    25, 50–51
By-products of trees, 31

California Desert
    Conservation Area
    (CDCA), 50

Carnivorous, 9, 40–41, 43,
    53
Center for Biological
    Diversity, 50
Conservation efforts, 14–17,
    24–25, 32–33, 34, 35,
    42, 46, 50–52
Coral reefs, 13
Crown, 30, 53

Death Valley, 45–46, 50
Deere, John, *23*
Desert, 44, 45–52

Ecosystem, 12, 13, 17, 31,
    35, 42, 43, 46, 50–52, 53
Edison, Thomas, 10
Emit, 8, 53
Endangered Species Act, 14
Endangered, 14, 53
Endemic, 16, 53

Everglades, *42*, 43
Exotic, 28, 53
Extinct, 8, 13–14, 17, 53

Feral, 33, 53
Fires, 22, 31, 34, 35
Forb, 19, 54
Forests, *26*, 27–35
Fort Irwin, 46, 48
Franklinia, *6*, 7–8
Fungus, 22, 24, 54

Grasses, *18*, 19–20
Gray, Asa, 16

Hawaiian Tree Cotton,
    33–34
Hawkmoth, 24
Hoover, Herbert, 46
Humans as exterminators,
    12–14, 17, 49–50

Indiangrass, 20

Jojoba, 10–11
Joshua trees, 46, *47*

Kauila, 34
Kaupulehu Forest Reserve,
    34, 35

King George County,
    Virginia, 32

Lane Mountain milkvetch,
    46
Leeward, 33, 54
Legume, 31, 54
Lichen, 49, 54
Lightning, 30
Longleaf Alliance, 32
Longleaf pines, 28–31
Loosestrife plant, 43

Meadowview Biological
    Research Station, 42
Medicinal value of plants, 8,
    9–10
Milkweed, 10, *11*
Mojave Desert, 45–52
Moth, 34
Mycorrhiza, 22, 54

National Marine Fisheries
    Service, 14
National Museum of Natural
    History, 8
National Park Service, 25
Native Americans, 21, 25
Nature Conservancy, 25,
    42

North Kona Dry Forest Working Group, 34
Novocain, 10, *10*

Old-growth, 30, 54
Orchids, 24

Pacific yew tree, 10
Pampas, 20
Peat, 38, *39*, 40, 54
Penetrate, 21, 54
Perennial, 24, 31, 54
Photosynthesis, 8, 55
Pitcher plants, 32, 40–41, *41*, 42
Plants' helpful purposes, 8–11, 33, 34
Plows, 23, 24
Pocosin, 38, 55
Pollinators, 11–12, 24
Prairie, *18*, 19–25, 55
Prairie fringed orchid, *2*, 24

Rain forests, 13, 27, 33
Red-cockaded woodpecker, 30, *30*

Shasta owl's clover, 16, *16*
Sierra Club, 50

Steppes, 20
Switchgrass, 20

Tallgrasses, 20, 23
Taproot, 28, 31, 55
Tar, 31
Taxol, 10
Threatened, 14, 55
Trees. *See* forests
Turpentine, 31

U.S. Congress, 14
U.S. Fish and Wildlife Service, 12, 14, 34, 50
U.S.D.A. Forest Service, 25

Veldts, 20
Venus flytrap, 41

Water vapor, 8
Wetlands, 36, 37–43
Wildflowers, 19, 22, 23
Wiliwili tree, 33
Willow tree, 10
Woodpecker, 30

Yellowstone National Park, *25*

About the Author

Reading about scientific developments and discoveries, exploring forests and seashores, observing wildlife, and introducing young people to the exciting world of nature are some of D. M. Souza's favorite things to do. As a freelancer, she has written more than two dozen science-related books, including *Freaky Flowers*, *Meat-Eating Plants*, *Plant Invaders*, *Wacky Trees*, and *What Is a Fungus?* for Franklin Watts.